NIST Technical Note 1630

Evaluation of Image Quality of Thermal Imagers used by the Fire Service

Francine Amon
Andrew Lock

NIST National Institute of Standards and Technology • U.S. Department of Commerce

NIST Technical Note 1630

Evaluation of Image Quality of Thermal Imagers used by the Fire Service

Francine Amon
Andrew Lock
Building and Fire Research Laboratory
National Institute of Standards and Technology

February, 2009

U.S. Department of Commerce
Carlos M. Gutierrez, Secretary

National Institute of Standards and Technology
Patrick D. Gallagher, Deputy Director

National Institute of Standards and Technology Technical Note 1630
Natl. Inst. Stand. Technol. Tech. Note 1630, 36 pages (February 2009)
CODEN: NSPUE2

ABSTRACT

This document reports on the test results of an evaluation of the image quality of thermal imaging cameras used by the fire service. The test methods used were consistent with the image quality test methods included in Draft National Fire Protection Association (NFPA) 1801 Standard on Thermal Imagers for the Fire Service. These tests measured the nonuniformity, spatial resolution, effective temperature range, and thermal sensitivity of fire service thermal imagers. Images used for these tests were collected using a high resolution visible camera focused on the thermal imager's display while the thermal imager viewed a variety of thermal targets. The laboratory test results were evaluated in terms of a multivariate model of human perception, which was based on tests conducted on firefighters viewing thermal images of representative scenes in which a fire hazard may be present.

1. INTRODUCTION

Five different fire service thermal imagers were evaluated by the National Institute of Standards and Technology (NIST) Fire Research Division using performance metrics and testing protocols that were developed specifically for evaluating thermal imaging cameras utilized by firefighters, many of which have subsequently been integrated into Draft NFPA 1801 Standard on Thermal Imagers for the Fire Service [1]. The performance metrics for these tests were Field of View (*FOV*), Nonuniformity (*NU*), Spatial Resolution (*SR*), Effective Temperature Range (*ETR*), and Thermal Sensitivity (*TS*). The thermal imagers tested were considered "black boxes" in the sense that a target was placed in the field of view and the resulting image that appeared on the thermal imager's display was captured by a high resolution Nikon D3[*] visible camera and processed using MATLAB software; the performance of the individual imaging components of the thermal imager was not measured. The methods of measurement and the results of these limited tests are presented in this report. The results of each individual test will be presented in the following five sections.

While these objective laboratory test methods were being developed, a parallel effort was made to investigate the appropriate pass/fail criteria to apply to them. Work was performed in collaboration with the US Army's Night Vision and Electronic Sensors Directorate (NVESD) in which the quality of thermal images common to firefighting applications was related to the ability of firefighters to perform a task. The task was to view a series of thermal images that had been degraded in specific ways and identify a potential fire hazard by clicking on it. The contrast (*C*), brightness (*B*), spatial resolution (*R*), and noise (*N*) of these images were each degraded to five different levels and the effects of these degraded factors on the test subject's ability to perform the task were analyzed separately and in combination. It was determined that the interactions between the *C*, *B*, *R*, and *N* factors were at least as important as these factors were individually to the firefighters' ability to identify fire hazards. In order to capture these important interactions in the pass/fail criteria, a multivariate model was created from the

---

[*] Certain companies and commercial properties are identified in this paper in order to specify adequately the source of information or of equipment used. Such identification does not imply endorsement or recommendation by the National Institute of Standards and Technology, nor does it imply that this source or equipment is the best available for the purpose.

perception test results and is used in Section 2.2 to evaluate the imaging performance of the thermal imagers. An image quality indicator (P) is introduced to predict the probability of successfully performing the fire hazard identification task based on the combination of image quality factors (C, B, R, and N) measured in the images collected from the thermal imagers undergoing laboratory tests. In this way, human perception is used to determine the quality of the image while the imagers are tested using objective test methods.

2. PERFORMANCE METRICS AND TEST METHODS

General specifications of the imagers tested are listed in Table 1. In some cases, imagers were not subjected to the full gamut of tests due to testing equipment failures. With respect to sensor type, the imagers have Amorphous Silicon (ASi) or Vanadium Oxide (VOx) heat sensing material.

Table 1. General specifications of thermal imagers tested.

Imager	Focal Plane Array	Sensor
1	320 x 240	ASi
2	320 x 240	VOx
3	160 x 120	ASi
4	320 x 240	VOx
5	80 x 60	VOx

2.1 Field of View
The field of view (FOV) for most of the thermal imagers was simply measured by placing a stiff plastic pointer along the centerline of the thermal imager's lens and positioning the camera such that the centerline of the outer surface of the lens was in the center of a protractor. The thermal imager was rotated until the edge of a thermal target placed 1 m away appeared to just touch the right edge of the displayed image. The angular position on the protractor indicated by the pointer was recorded. The thermal imager was then rotated about the center of the protractor until the same edge of the thermal target appeared to just touch the left edge of the displayed image. The angular position on the protractor indicated by the pointer was again recorded. The horizontal FOV is equal to the sum of the left and right circumscribed angles. This process was performed three times for each imager and averaged to determine the horizontal FOV. The same procedure was conducted with the thermal imager rotated 90 degrees about the axis of its optical system to determine the vertical FOV. For Imager 2, which was tested at a later date than the other imagers, a rotating optical platform was used in the same way as described above, using angular markings inscribed on the platform rather than the plastic pointer. The results are provided in Table 2.

The FOV was not included in the image quality perception tests conducted at NVSED because it is simply a measurement of the viewing angle of the thermal imager and thus does not directly affect image quality. Indirectly, image quality is related to the level of detail that is discernable

2

within the field of view, and this is captured by measuring spatial resolution, which is an important factor in image quality and is measured in both the laboratory tests and the human perception tests. Also, the *FOV* does affect firefighter performance in that a narrower (versus

Table 2. Horizontal and vertical *FOV* results.

Imager	Vertical *FOV* (degrees)	Horizontal *FOV* (degrees)	Aspect Ratio
1	33 ± 0.6	42.5 ± 0.6	0.78 to 1
2	28 ± 0.6	33.5 ± 0.6	0.84 to 1
3	33 ± 0.6	42.5 ± 0.6	0.78 to 1
4	32 ± 0.6	43.0 ± 0.6	0.74 to 1
5	19 ± 0.6	24.0 ± 0.6	0.79 to 1
The uncertainty expressed in these test measurements is a combination of Type A (statistical) and Type B (other), with a coverage factor of 2 resulting in a 95 % confidence interval. See Section 5 for detailed uncertainty analysis.			

wider) *FOV* limits the situational awareness of a firefighter that uses the imager to scan or navigate a smoke-filled room.

The information provided in Table 2 shows that the design of thermal imager optical systems varies significantly, even in cases where the same sensing material and detector array size are used. The percent difference in average (combined vertical and horizontal) *FOV* between Imager 1 and Imager 5 is 75 %.

2.2 Nonuniformity

Nonuniformity (*NU*) is a measure of the thermal imager's response to a uniform thermal target [2, 3]. A common malady of thermal imaging is the phenomenon whereby the individual subcomponents (pixels) of the thermal detector drift away from their initial values over time due to minor variations in the sensing material and electronic circuitry. To compensate for this drift, some types of imagers employ a shutter that closes over the detector and momentarily blocks incoming radiation, effectively resetting all the pixels to the ambient temperature of the shutter. The shutter is activated according to proprietary algorithms; this process is termed a nonuniformity correction (NUC).

Three tests for each thermal imager were conducted at each of five target temperatures: 1 °C, 30 °C, 100 °C, 160 °C, and 260 °C, which span the temperature range of interest to the fire service [4]. Well-characterized extended-area blackbodies were used as targets (a CI Systems SR800 was used for target temperatures below 260 °C and a CI Systems SR80-HT was used for the 260 °C target temperature). The thermal imager under test was positioned such that the image of the blackbody surface completely filled the field of view and the thermal imager was normal to the plane of the blackbody surface.

A Nikon D3 digital visible camera was used to take ten 16-bit digital photos of the images displayed by the thermal imager under test for each target temperature. These images were

3

initially stored in flash memory on the Nikon, and then transferred to a PC for further processing. The processing steps were as follows:

1. Convert the images from the proprietary Nikon NEF format to 16-bit TIFF files.

2. Convert the TIFF files to grayscale as defined per IEC 61966-2-1 (1999) [5].

3. Use a MATLAB program to define a region of interest (ROI) that encompasses at least 90 % of the FOV of the thermal imager under test. Exclude or remove symbols, icons, and text from this ROI.

4. Filter out the high-frequency noise created by oversampling the thermal imager's display. This was accomplished with an averaging filter which averaged over an area whose width and height was determined by the frequency of the noise.

5. Calculate the *NU* for each image using the following equation [3]

$$NU = \frac{\sigma}{\mu} \tag{1}$$

Where σ is the standard deviation and μ is the mean, respectively, of pixel intensity values in the ROI. Retain the original pixel intensity values for later use.

6. Rank the ten *NU* values collected at each target temperature, then discard the highest *NU* value from the data set. This step is performed to mitigate the affect of an ill-timed NUC that may occur during the collection of images.

7. Average the *NU* values for the three individual tests conducted at each target temperature. This is the *NU* value reported for each target temperature in Table 3.

The accuracy of the temperature measurements described in this section is discussed in Section 5, uncertainty analysis.

Table 3. *NU* values for a range of set point temperatures.

Imager	*NU* 1 °C	*NU* 30 °C	*NU* 100 °C	*NU* 160 °C	*NU* 260 °C
1	0.324 ± 0.019	0.497 ± 0.049	0.619 ± 0.021	0.471 ± 0.010	0.161 ± 0.100
2	0.457 ± 0.023	0.360 ± 0.046	0.281 ± 0.013	0.296 ± 0.005	0.235 ± 0.022
3	0.395 ± 0.145	0.393 ± 0.043	0.544 ± 0.007	0.702 ± 0.038	0.665 ± 0.013
4	0.524 ± 0.042	0.355 ± 0.096	0.617 ± 0.024	0.172 ± 0.016	0.761 ± 0.038
5	0.239 ± 0.007	0.339 ± 0.004	0.397 ± 0.032	0.404 ± 0.214	0.349 ± 0.006
The uncertainty expressed in these test measurements is a combination of Type A (statistical) and Type B (other), with a coverage factor of 2 resulting in a 95 % confidence interval. See Section 5 for detailed uncertainty analysis.					

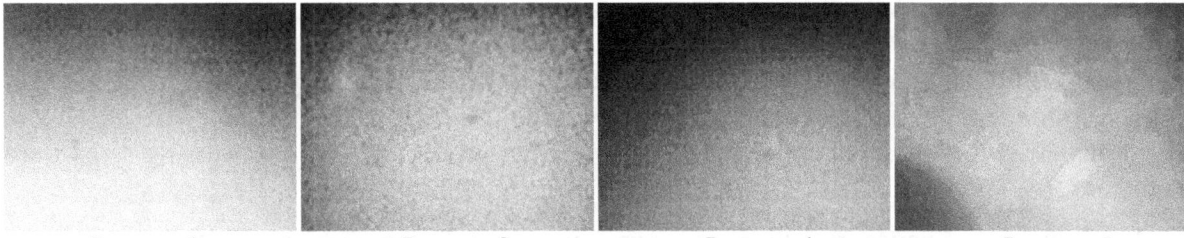

| Imager 1 | Imager 3 | Imager 4 | Imager 5 |

Figure 1. Representative images used for laboratory measurements of nonuniformity. The target temperature is 100 °C.

A representative image from each of the thermal imagers, taken at the target temperature of 100 °C is shown in Figure 1. Note that these images have been modified to remove identifying symbols and icons that would normally reside within the displayed imaged.

2.3 Spatial Resolution

The spatial resolution (*SR*) performance metric is a measure of the ability of firefighter thermal imagers to reproduce the details of a scene or target [6]. The spatial resolution test is used both to measure an imager's spatial resolution and to determine whether a design robustness test, e.g., the corrosion resistance test, has impacted the imager's ability to produce images of acceptable quality.

This test was performed three times for each imager and the results were averaged to determine the *SR*. Similar to the nonuniformity test, the *SR* was measured by viewing an image on the thermal imager's display with a high resolution digital visible camera. The thermal imager viewed a thermal target comprised of two sets of converging lines, as shown in Figure 2. This target is a portion of the target used by the ISO Spatial Resolution Standard ISO 12233 [7]. The target foreground and background were coated with well-characterized black paint having an emissivity of 0.94 ± 0.05. The characterization of the black paint was performed by the paint manufacturer.

Figure 2. Spatial resolution thermal target. The foreground (black markings) is held at a constant temperature of 3 °C above ambient. The target size is 61 cm measured along the centerline of each of the two converging line sets.

The thermal imager under test was placed 1 m from the target, normal to the plane of the target, and oriented to focus on the center of the target. A Nikon D3 digital visible camera was used to take ten 16-bit digital photos of the images displayed by the thermal imager under test, as shown in the left image in Figure 3. These images were initially stored in flash memory on the Nikon, then transferred to a personal computer (PC) for further processing. The processing steps were as follows:

1. Convert the images from the proprietary Nikon NEF format to 16-bit TIFF files.

2. Convert the TIFF files to grayscale as defined per IEC 61966-2-1 (1999) [5].

3. Filter out the high-frequency noise created by oversampling the thermal imager's display. This was accomplished with an averaging filter which averaged over an area whose width and height was determined by the frequency of the noise.

4. Use a MATLAB program to rotate the image 45 degrees and select an ROI that encompasses the converging lines, as shown in the right image in Figure 3, and calculate the contrast transfer function (*CTF*) at least at each of the index numbers along the two sets of converging lines; a process that calculates the CTF at each row in the ROI is also acceptable. The *CTF* is calculated using the following equation:

$$CTF = \frac{I_{max} - I_{min}}{I_{max} + I_{min}} \tag{2}$$

where I_{max} and I_{min} are the highest and lowest pixel intensity values, respectively, along a row of pixels cut through the pattern at least at each index line as indicated by the dotted lines in the right image in Figure 3. Retain the original pixel intensity values for later use. Pixels that represent symbols, icons, or text are excluded from the analysis.

5. The *CTF* values are multiplied by $\pi/4$ and paired with the frequencies of each index number to construct a Modulation Transfer Function (*MTF*) curve. The frequencies are listed in Table 4 below.

Table 4. Frequencies corresponding to the indices on the spatial resolution target when the thermal imager under test is placed 1 m from the target.

Index	Frequency (cycles/mrad)
1 (largest end of lines)	0.019
2	0.038
3	0.083
4	0.118
5	0.143
6	0.167
7	0.200
8	0.250
9	0.286
10 (smallest end of lines)	0.300

6. The *MTF* curve is normalized to the value obtained at the low-frequency edge of the target (index 1).

7. The average *NU* value measured at a target temperature of 30 °C, and listed in Table 3, is subtracted from the *MTF* curve. This adjustment corrects for the effects of high frequency noise.

8. The area under the normalized and adjusted *MTF* curve is the spatial resolution.

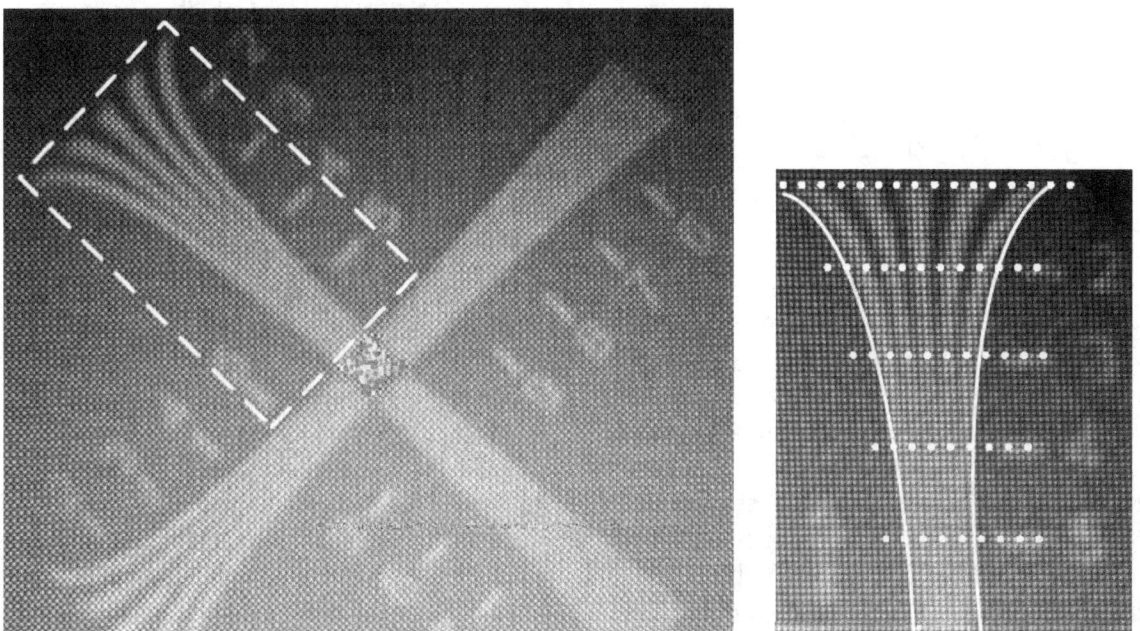

Figure 3. Spatial resolution image on left with a dashed box indicating the inset image on right. The ROI is between the solid white lines in the right image and the dotted lines show the position and orientation of the rows of pixels used to calculate the *CTF*.

The accuracy of the temperature measurements described in this section is discussed in Section 5, uncertainty analysis.

Table 5. Spatial resolution test results.

Imager	*SR*
1	0.0442 ± 0.0039
2	0.0389 ± 0.0043
3	0.0305 ± 0.0008
4	0.0402 ± 0.0089
5	0.0445 ± 0.0013
The uncertainty expressed in these test measurements is a combination of Type A (statistical) and Type B (other), with a coverage factor of 2 resulting in a 95 % confidence interval. See Section 5 for detailed uncertainty analysis.	

2.4 Effective Temperature Range

The Effective Temperature Range (*ETR*) test measures the ability of a firefighter thermal imager to see relatively small temperature differences in cases when large temperature differences exist in the field of view [1]. In this test, the thermal imager is positioned such that it views a set of contrast bars having constant temperature while the temperature of a surface of equal size in the field of view is increased from near ambient to 550 °C. In general, as the hot surface temperature increases, the contrast of the bars decreases. This test, as described in Draft NFPA 1801, also has a color component in which the colorization corresponding to certain surface temperature ranges is verified. The colorization employed by the imagers tested for this work was not designed to conform with Draft NFPA 1801, therefore the color component of this test is not reported here.

The thermal imagers were placed 1 m from the bar target, which were comprised of four vertical 1.27 cm diameter copper tubes placed 1.27 cm apart, resulting in a bar frequency of 0.04 cycles/mrad. A heated siloxane solution flowed through the bars to maintain a temperature of 37 °C ± 1 °C. The background temperature was 28 °C ± 1 °C. The bars and background were coated with black paint having an emissivity of 0.94 ± 0.05. A CI Systems SR80-HT extended area blackbody having a 178 mm square surface was positioned such that its radiation impinged on a 203 mm diameter off-axis parabolic mirror having a focal length of 1 m and an offset of 10 degrees and was directed to the thermal imager under test. The layout of the testing equipment is shown in Figure 4. It is important that the emitting surface of the blackbody appear in the center of the image displayed by the thermal imager, while the heated bars appear at one side.

The size of the blackbody radiation in the thermal imager's field of view was at least as large as the bars, but varied depending on the field of view of the thermal imager under test. A Nikon D3 digital visible camera was used to take 16-bit digital photos of the image displayed by the thermal imager every 3 seconds. These images were initially stored in flash memory on the Nikon, then transferred to a PC for further processing. The processing steps were as follows:

1. Convert the images from the proprietary Nikon NEF format to 16-bit TIFF files.

2. Convert the TIFF files to grayscale as defined per International Electrotechnical Commission (IEC) 61966-2-1 (1999) [5].

3. Filter out the high-frequency noise created by oversampling the thermal imager's display. This was accomplished with an averaging filter which averaged over an area whose width and height was determined by the frequency of the noise.

4. Use a MATLAB program to define an ROI that encompasses the four vertical bars. Exclude or remove symbols, icons, and text from this ROI. Retain the original pixel intensity values for later use.

5. Use the MATLAB program to calculate the *CTF* of the bars within the ROI. The *CTF* is calculated using equation 2.

6. The *ETR* is the temperature at which the *CTF* of the bars falls below a predetermined value. The cut-off value is not specified for this test in lieu of the image quality indicator discussed in Section 2.2.

The *ETR* data is plotted and discussed in Section 3.4 in the context of the image quality indicator.

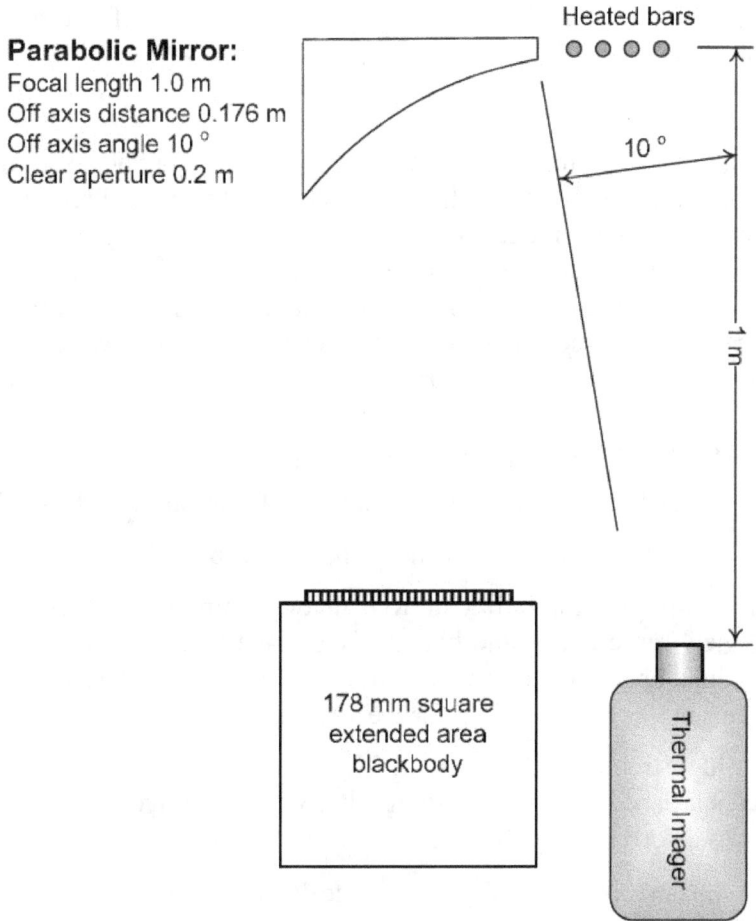

Figure 4. *ETR* testing configuration. The target consists of the heated bars, which appear on the side of the image, and the emitting surface of the blackbody, which appears in the center of the image.

9

2.5 Thermal Sensitivity

The thermal sensitivity (*TS*) test measures the ability of fire fighting thermal imagers to discern small temperature differences [1]. Three tests for each thermal imager were conducted at each of five target temperatures: 1 °C, 30 °C, 100 °C, 160 °C, and 260 °C, which span the temperature range of interest to the fire service [4]. In most cases, a pair of well-characterized extended area blackbodies were used as targets. A CI Systems SR800 was used for target temperatures between 1 °C and 160 °C, which is the temperature range of this blackbody, a CI Systems SR80-HT was used for target temperatures above 30 °C since this blackbody cannot be cooled below 30 °C, and an IRCon BCN-07C cavity blackbody combined with a 20.3 cm diameter spherical mirror having a focal length of 30.5 cm was used for the 260 °C target temperature because the SR800 could not produce this temperature. A metal 3.8 liter (1 gallon) container coated with black paint having an emissivity of 0.94 ± 0.05 filled with ice water was used for the 1 °C target temperature because only the SR800 could be cooled below ambient temperatures. The thermal imager under test was positioned such that the image of two blackbody surfaces equally filled as much of the field of view as possible. The amount of the field of view filled by the blackbody surfaces depended on the optical geometry of the thermal imager under test. The thermal imager was positioned normal to the plane of the blackbody surfaces. The purpose of having two blackbody surfaces in the field of view was to force the automatic gain control (AGC) function of the imagers to stabilize during the test. A significant portion of a typical image used to calculate *TS* consists of surfaces at ambient temperature, which also contributed to the stabilization of the AGC.

For each nominal target temperature (T_n) below 260 °C, one blackbody was held at a constant temperature. The other blackbody was initially set at $T_n - 0.05$ °C and ten 16-bit digital photos were collected from the thermal imager display with a Nikon D3 digital visible camera. The blackbody was then set at T_n and ten more images were collected. The blackbody was then set at $T_n + 0.05$ °C and ten more images were collected. For the 260 °C nominal target temperature, the same process was performed except that the temperature differences were $T_n \pm 0.20$ °C. This entire process was conducted three times for each imager at each nominal target temperature. The images were initially stored in flash memory on the Nikon, then transferred to a PC for further processing. The processing steps were as follows:

1. Convert the images from the proprietary Nikon NEF format to 16-bit TIFF files.

2. Convert the TIFF files to grayscale as defined per IEC 61966-2-1 (1999) [5].

3. Use a MATLAB program to define an ROI that encompasses at least 90 % of the area in the field of view representing the blackbody surface that is set at $T_n \pm 0.05$ °C or $T_n \pm 0.20$ °C. Exclude or remove symbols, icons, and text from this ROI. The three images in Figure 5 are examples of ROIs used for the *TS* test.

4. Filter out the high-frequency noise created by oversampling the thermal imager's display. This was accomplished with an averaging filter which averaged over an area whose width and height was determined by the frequency of the noise.

5. Calculate the mean pixel intensity (μ) of the ROI. Retain the original pixel intensity values for later use.

6. Rank the ten μ values collected at each target temperature, then discard the lowest μ value from the data set. This step is performed to mitigate the affect of an ill-timed NUC that may occur during the collection of images.

7. Average the remaining nine μ values for each set of images collected at each nominal target temperature, i.e., at T_n and $T_n \pm 0.05$ °C or $T_n \pm 0.20$ °C.

8. Calculate the contrast between the average pixel intensity at T_n and $T_n - 0.05$ °C or 0.20 °C. Record this value. Calculate the contrast between the average pixel intensity at T_n and $T_n + 0.05$ °C or 0.20 °C. Record this value. Average the two contrast values.

9. Average the contrast values per step 8 above for the three individual tests conducted at each target temperature. This is the thermal sensitivity value reported for each target temperature in Table 6.

Table 6. Thermal sensitivity test results.

Imager	*TS* 1 °C	*TS* 30 °C	*TS* 100 °C	*TS* 160 °C	*TS* 260 °C
1	0.0134 ± 0.0123	0.0029 ± 0.0012	0.0028 ± 0.0018	0.0042 ± 0.0006	0.0096 ± 0.0010
2	0.0316 ± 0.0213	0.0070 ± 0.0029	0.0030 ± 0.0015	0.0021 ± 0.0011	0.0011 ± 0.0003
3	0.0096 ± 0.0047	0.0067 ± 0.0023	0.0020 ± 0.0009	0.0020 ± 0.0020	0.0092 ± 0.0107
4	0.0282 ± 0.0110	0.0042 ± 0.0007	0.0016 ± 0.0002	0.0041 ± 0.0015	0.0026 ± 0.0009
5	0.0237 ± 0.0183	0.0104 ± 0.0039	0.0039 ± 0.0027	0.0025 ± 0.0007	0.0063 ± 0.0012

The uncertainty expressed in these test measurements is a combination of Type A (statistical) and Type B (other), with a coverage factor of 2 resulting in a 95 % confidence interval. See Section 5 for detailed uncertainty analysis.

An example of the contrast obtained from a *TS* test is presented in Figure 5. Note that, while there may be a discernable difference in the average pixel intensity of the three images, there may not be enough of a difference to provide useful images for the user of the thermal imager. The very low contrasts recorded for the *TS* tests may not be sufficient for firefighters to locate fire hazards or perform other tasks, so additional tests examined the change in contrast as the value used for T_n was increased. Three tests were conducted at each T_n value for one imager. The results are shown in Figure 6 and will be discussed in Section 3.4 in the context of the image quality indicator.

Figure 5. Example test images for *TS* at a nominal target temperature of 100 °C. The image on the left is of a 100.05 °C target, the image in the center is of a 100.00 °C target, and the image on the right is of a 99.95 °C target. Note that these three images may appear to be identical when viewed on a printed page.

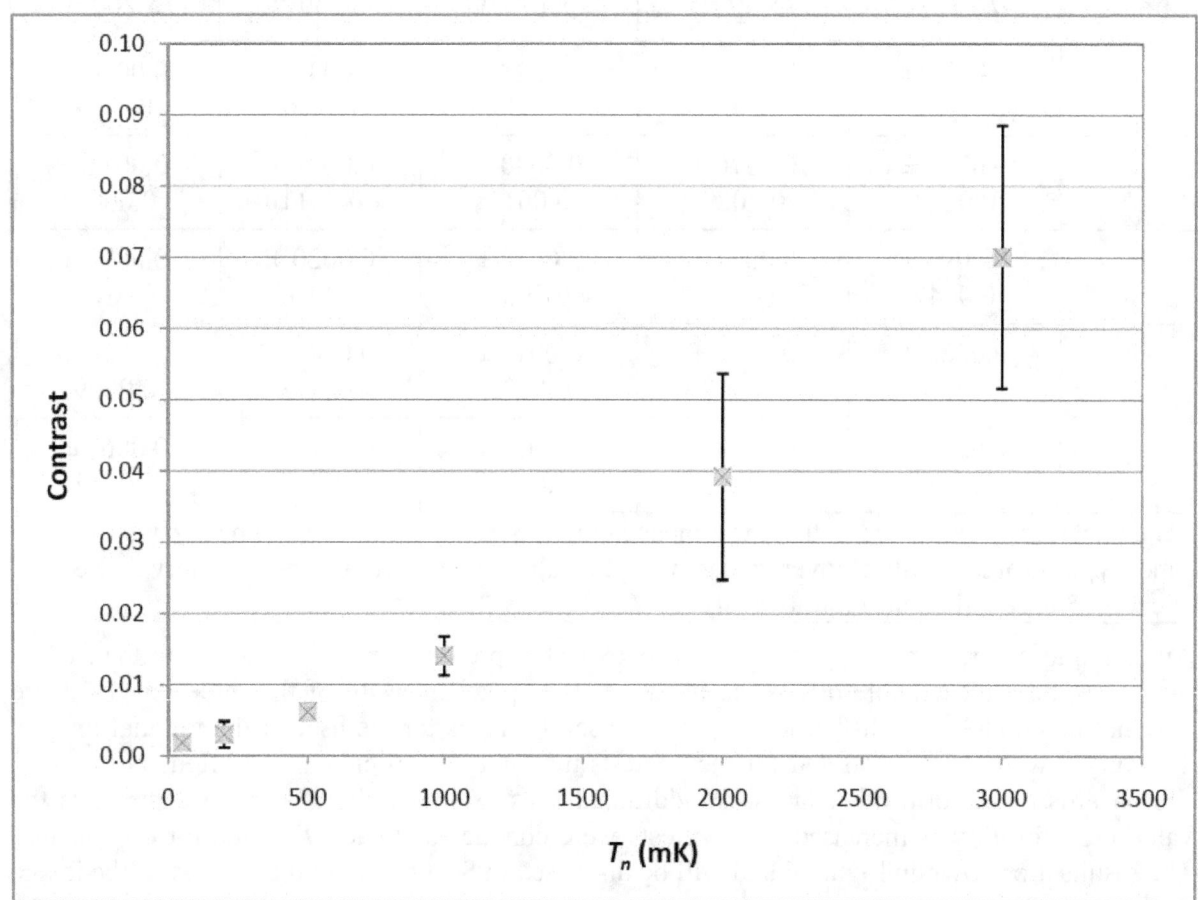

Figure 6. Contrast response to increasing values of T_n for the *TS* test. The nominal target temperature was 100 °C. The uncertainty expressed in these test measurements is a combination of Type A (statistical) and Type B (other), with a coverage factor of 2 resulting in a 95 % confidence interval. See Section 5 for detailed uncertainty analysis.

12

3. MULTIVARIATE ANALYSIS

In order to determine the minimum level of image quality that allows users to successfully perform meaningful tasks, a set of perception tests were conducted using firefighters that use thermal imagers on the job. The test subjects were asked to observe thermal images of scenes in which a single potential fire hazard may or may not be present. The test subjects identified the hazard by clicking on it or, if no hazard was present, the test subject clicked a "No Hazard" button. The contrast (C), brightness (B), spatial resolution (R), and noise (N) in the thermal images were degraded in varying degrees such that the effect of each of these primary quality factors on the ability of a firefighter to perform a task was quantified. These primary factors were chosen based on their relevance to the tests presented in Sections 2.2 to 2.5, although there is not a 1-to-1 correspondence. The measurement of contrast is fundamental to all the imaging tests conducted for this report. Brightness is used as an indication of the relative temperature of surfaces in the field of view. Broadband noise, manifested as nonuniformity, and spatial resolution are directly related to the test methods described in Sections 2.2 and 2.3, respectively.

During the process of analyzing the perception test results, it became apparent that the above four primary factors of image quality were too tightly dependent upon one another to adequately describe the imaging performance of a thermal imager if used in separate, individual performance tests such as those presented in Sections 2.2 to 2.5. For example, the interaction of contrast with brightness has a much more profound effect on a firefighter's ability to successfully identify a fire hazard than the additive effect of contrast plus brightness. In order to accommodate the interactions between primary image quality factors without disturbing the test methods previously discussed, a human perception model was developed that includes the four primary factors, as measured by the test methods described in Sections 2.2 to 2.5, along with two-way interactions between these factors and squared terms of these factors. The model is presented in its most compact form in the following equation:

$$P = \frac{e^X}{1 + e^X} \tag{3}$$

where P is the probability of successfully identifying the fire hazard, or absence of a fire hazard, in an image. For the purposes of this report, P is the image quality indicator. The variable X represents a long set of terms in the form

$$X = 0.2563 + C + B + R + N + C^2 + B^2 + R^2 + N^2 + CB + CR + CN + BR + BN + RN \tag{4}$$

where C, B, R, and N are the contrast, brightness, spatial resolution, and noise that were present in the images observed by the firefighters during their perception tests. The coefficients assigned to each of the terms in equation 4 are listed in Table 7 below.

Table 7. List of coefficients that apply to terms in equation 4.

Coefficient	Term
0.2563	intercept
0.8737	C
0.4595	B
15.85	R
-2.567	N
-8.359	C^2
-4.631	B^2
-242.2	R^2
2.893	N^2
3.779	CB
48.71	CR
11.60	CN
34.91	BR
-8.016	BN
-5.008	RN

Several methods were used to validate the model represented by this equation, which will be discussed in depth in a future publication. One of the methods used is to plot the averaged data collected in the perception tests for each of the 25 different image degradation settings against values predicted by the model. This is done in Figure 7, which shows a very good correlation between actual data and predicted values.

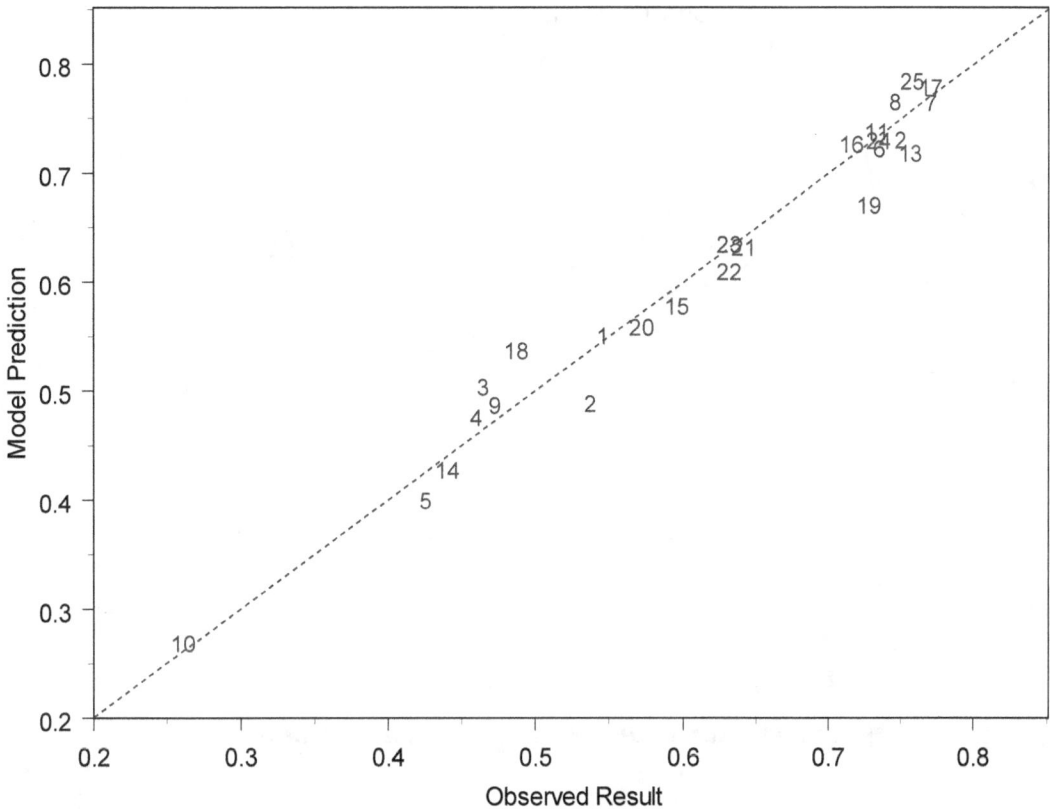

Figure 7. A comparison of the observed and model predicted values for the probability of successfully completing the fire hazard identification task. The plot character represents the image setting (1 to 25). A perfect correlation falls at the 45° diagonal.

Finally, a consistency check was performed to observe the predicted values from the model as each primary image quality factor was varied across the range of possible values, holding the other factors constant at their average value. This is shown in Figure 8. Note that all the factors vary from 0 to 1 except R, which is plotted using the uppermost x-axis. Also, note that the model is useful for predictions within the range of factor values that were used in the perception tests; extrapolation beyond this range of values for each factor is less useful.

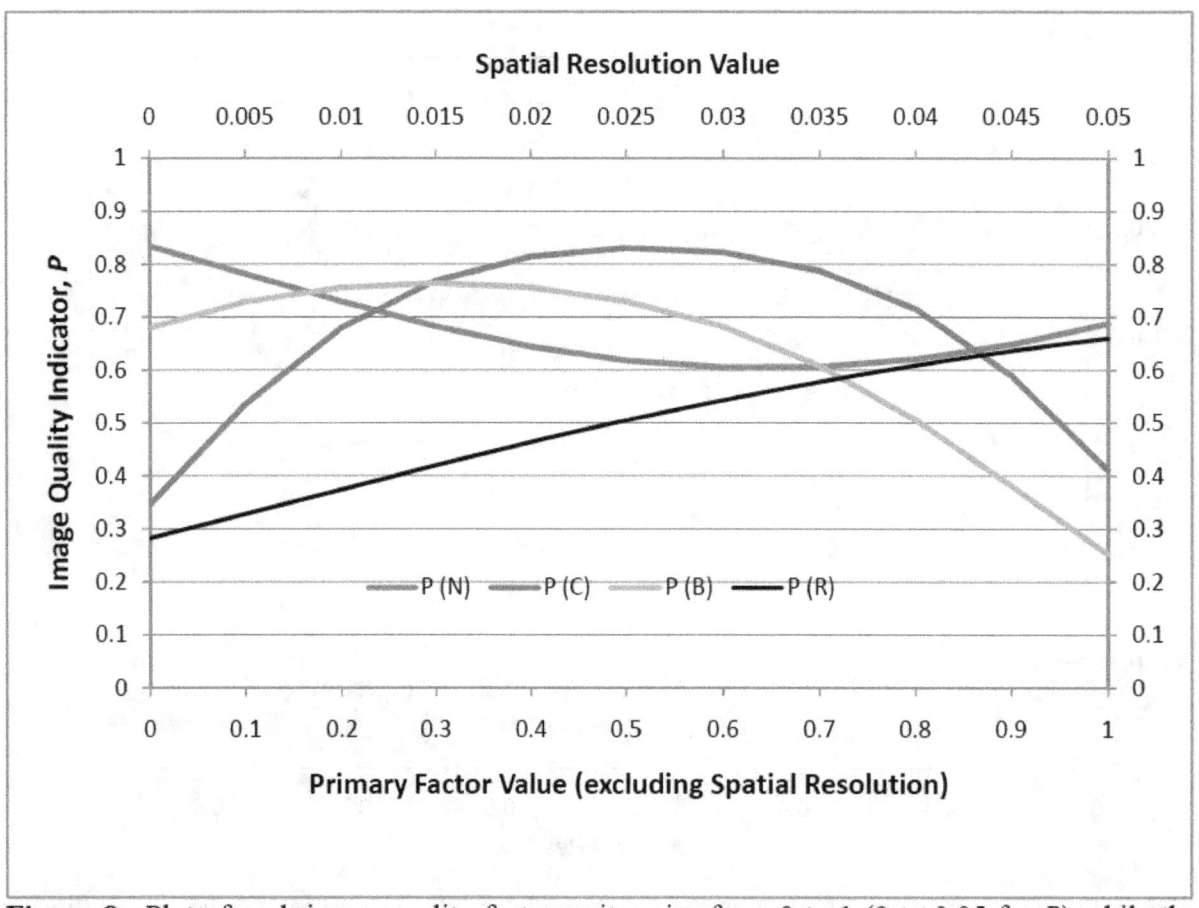

Figure 8. Plot of each image quality factor as it varies from 0 to 1 (0 to 0.05 for R) while the other factors are held constant at their average values.

In order to optimize the perception testing procedure and reduce test subject fatigue, the range of values used for the primary image quality factors in the degraded images were chosen to explore a particular region of the probability of successfully performing the fire hazard identification task, this region was $0.6 < P < 0.9$. The reasoning was that collecting data from images in which none of the test subjects were able to identify anything did not yield valuable information for the purposes of this work. Likewise, using images in which all test subjects were able to identify the fire hazard would not provide useful information from which to construct the model. Therefore, no images were degraded to "zero" values of any of the primary image quality factors. It should be also noted that the upper end of the N primary factor was not within the range of N values used to construct the model and values above about 0.6 are extrapolated.

3.1 Nonuniformity and Spatial Resolution Procedure
The measurement of nonuniformity at the target temperature of 30 °C is independent of the image quality equation and is performed strictly in conformance with the method described in Section 2.2, in which the standard deviation of the pixel intensity is divided by the mean pixel intensity of a ROI within an image. This is the only test that does not require the use of equation 3 and is used to establish the primary factor R. The nonuniformity tests conducted at target temperatures of 1 °C, 100 °C, 160 °C, and 260 °C make use of equation 3 and therefore

necessitate a pre-existing value for *R*. This will become apparent as the procedure is explained below.

1. For the target temperature of 30 °C, follow the *NU* procedure outlined in Section 2.2.

2. Follow the procedure outlined in Section 2.3 to calculate the spatial resolution, *SR*. To apply equation 3, use the contrast measured at the low-frequency edge of the spatial resolution target as *C*, use the average pixel intensity within the spatial resolution ROI as *B*, use the measured spatial resolution as *R*, and use the nonuniformity measured in step 1 above as *N*.

3. Calculate the image quality indicator, *P*. If *P* indicates less than the desired image quality level, then the imager fails the entire test sequence and no further testing is necessary.

4. If *P* indicates image quality that is greater than or equal to the desired level at the target temperature of 30 °C, apply equation 3 to the nonuniformity data collected at target temperatures of 1 °C, 100 °C, 160 °C, and 260 °C, applying equation 3 separately to data collected at each target temperature. For these tests, use the same contrast value as in step 2 for *C*, use the average pixel intensity within the nonuniformity ROI as *B*, use the same measured spatial resolution as in step 2 for *R*, and use the nonuniformity measured according to the procedure outlined in Section 2.2 at each of the target temperatures (1 °C, 100 °C, 160 °C, and 260 °C) as *N*.

5. Calculate the image quality indicator, *P*. If *P* indicates less than the desired image quality level, then the imager fails the entire test sequence and no further testing is necessary.

6. If *P* indicates image quality that is consistently greater than or equal to the desired level, continue to the test sequence for effective temperature range in the following section.

3.2 Effective Temperature Range Procedure

The procedure for applying equation 3 to the data collected for the effective temperature range test is relatively simple. For this test, the image quality indicator, *P*, is calculated using pixel intensities from the ROI encompassing the bars for each image collected during the test. The procedure is explained below.

1. Follow the *ETR* procedure outlined in Section 2.4.

2. To apply equation 3, use the *CTF* of the bars as the contrast *C*, use the average pixel intensity of the bar ROI as the brightness *B*, use the same measured spatial resolution as in step 2 of Section 3.1 for *R*, and use the nonuniformity value obtained at a target temperature of 30 °C in step 1 of Section 3.1 as *N*.

3. Calculate the image quality indicator, *P*, for each image collected during the test. If any of the calculated values of *P* indicate less than the desired image quality level, the imager under test fails the entire test sequence and no further testing is necessary.

4. If *P* indicates image quality that is consistently greater than or equal to the desired level, continue to the test sequence for thermal sensitivity in the following section.

3.3 Thermal Sensitivity Procedure

The procedure for applying equation 3 to the data collected for the thermal sensitivity test is also relatively simple. For this test, the image quality indicator, P, is calculated using a contrast measurement that depends on subtle changes in pixel intensity resulting from small changes in the target surface temperature. The procedure is explained below.

1. Follow the *TS* procedure in Section 2.5.

2. To apply equation 3, for each nominal target temperature, use the contrast obtained in step 9 of Section 2.5 as C, use the average pixel intensity of the bar ROI as the brightness B, use the same measured spatial resolution as in step 2 of Section 3.1 for R, and use the nonuniformity value obtained at the corresponding target temperature as N.

3. Calculate the image quality indicator, P, at each nominal target temperature. If any of the calculated values of P indicate less than the desired image quality level, the imager under test fails the entire test sequence and no further testing is necessary.

4. If P indicates image quality that is consistently greater than or equal to the desired level, the imager under test has successfully completed the sequence of image quality tests.

3.4 Multivariate Image Quality Results

The results shown in this section consider the same test conditions and data that were collected using the test methods presented in Sections 2.2 to 2.5 but offer the added benefit of incorporating the effects of interactions between the primary image quality factors. Using equation 3 in this way allows the pass/fail criteria for the image quality test methods to be based on human perception of images that relate to tasks performed in the user's line of duty.

3.4.1 Nonuniformity

The nonuniformity (*NU*) results, expressed in terms of the image quality indicator, P, are shown below in Figure 9. The mid-range target temperatures of 100 $^{\circ}$C and 160 $^{\circ}$C, at which some of the imagers show a larger uncertainty, are in the vicinity of the mode-shift trigger for those imagers. An automatic mode shift is a method used in thermal imagers that employ a microbolometer type detector to extend the dynamic range of the imager by decreasing the integration time that the thermal scene is exposed to the detector array. A side effect of operating at the shorter integration time is reduced thermal sensitivity; this will be explained in more detail in the thermal sensitivity discussion. The effect on the nonuniformity results presented here is that the imager may automatically shift for one of the three tests conducted at each target temperature, and not shift for the other two tests if the mode shift algorithm isn't triggered. The mode shift algorithm is a proprietary process and can depend on many conditions in the thermal scene, including but not limited to target temperature. This phenomenon can make testing the imagers problematic if the target temperature and other conditions are very close to the mode shift conditions.

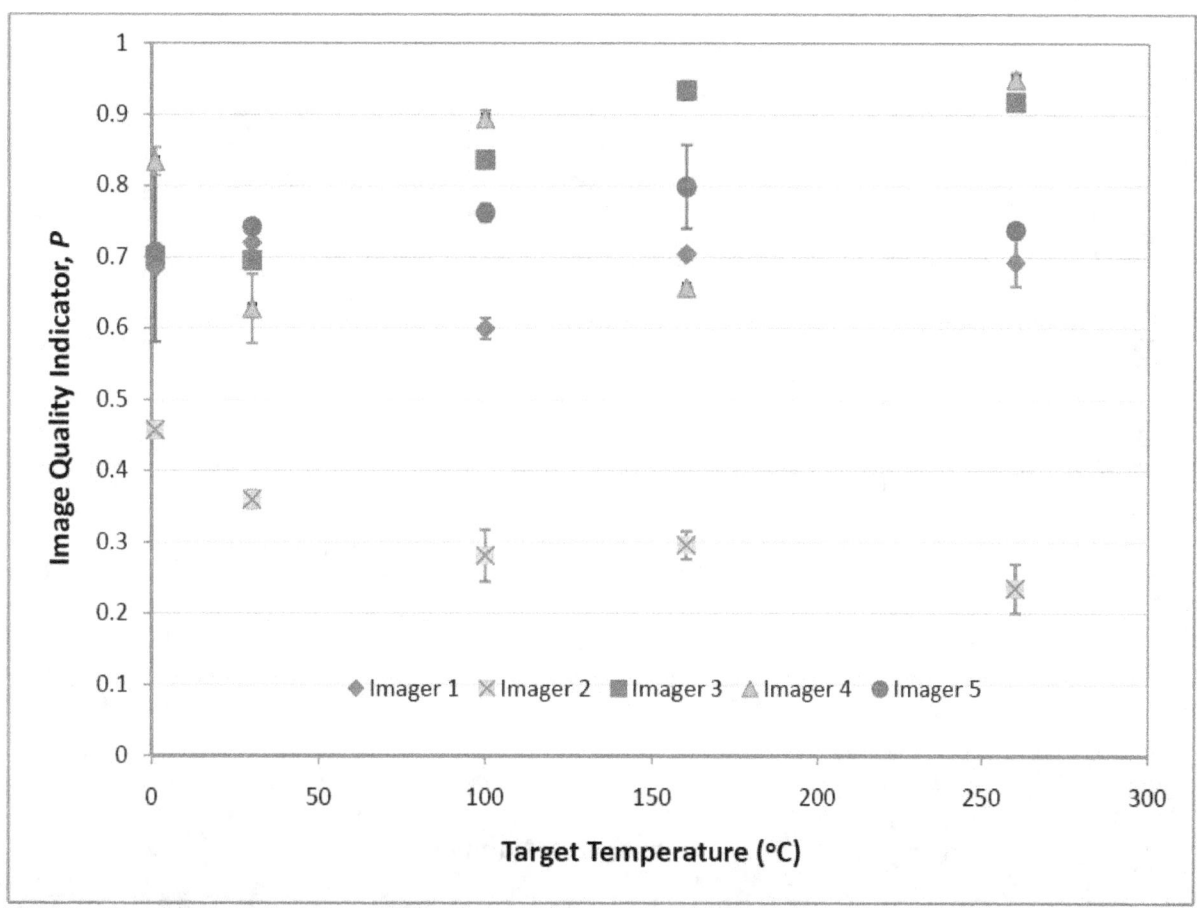

Figure 9. Nonuniformity (*NU*) plotted in terms of the image quality indicator, *P*. The uncertainty expressed in these test measurements is a combination of Type A (statistical) and Type B (other), with a coverage factor of 2 resulting in a 95 % confidence interval. See Section 5 for detailed uncertainty analysis.

3.4.2 Spatial Resolution

The spatial resolution (*SR*) results, expressed in terms of the image quality indicator, *P*, are shown below in Figure 10. With the exception of imager 1, the imagers produced similar *SR* results in spite of having different values of *C*, *B*, *R*, and *N* in equation 3.

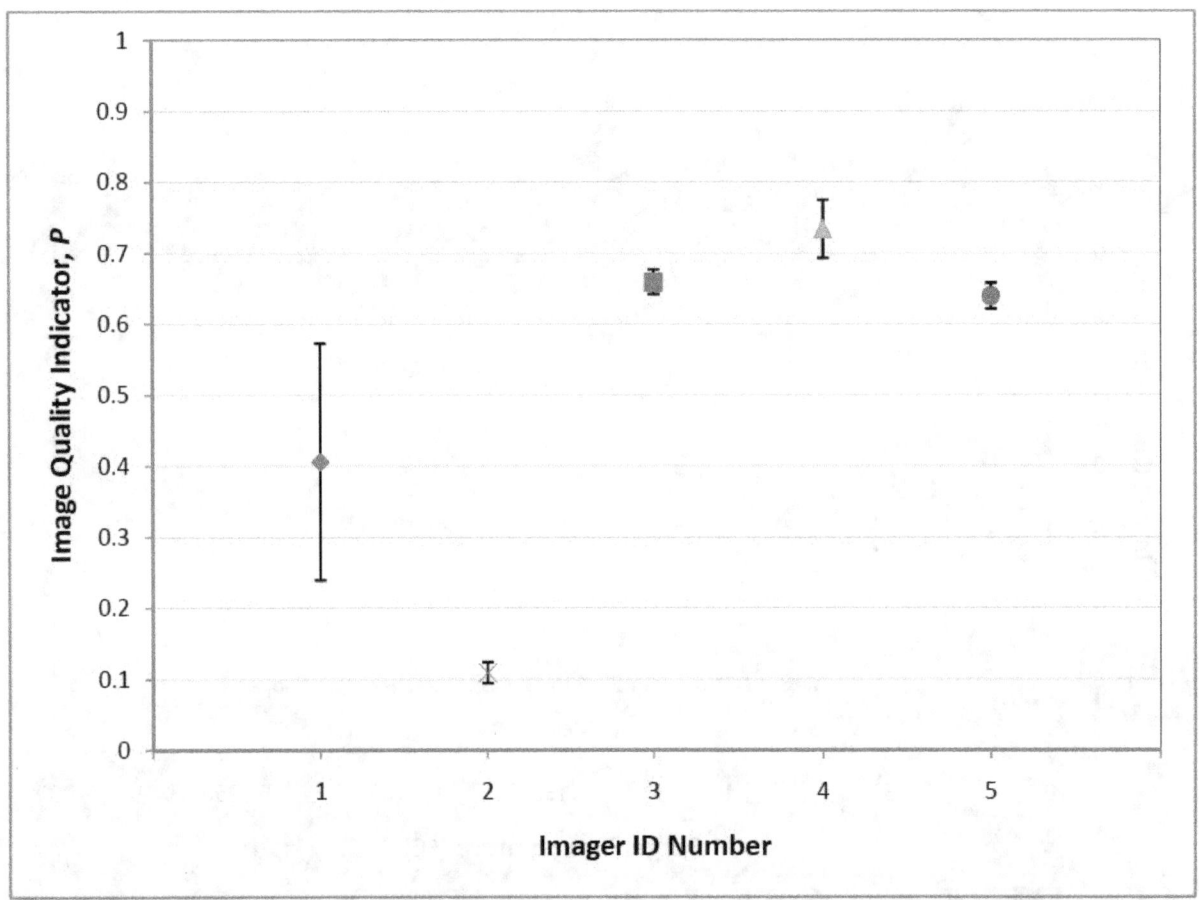

Figure 10. Spatial resolution (*SR*) plotted in terms of the image quality indicator, *P*. The uncertainty expressed in these test measurements is a combination of Type A (statistical) and Type B (other), with a coverage factor of 2 resulting in a 95 % confidence interval. See Section 5 for detailed uncertainty analysis.

3.4.3 Effective Temperature Range

The effective temperature range (*ETR*) results for Imagers 3, 4, and 5, expressed in terms of the image quality indicator, *P*, are shown below in Figures 11, 12, and 13, respectively. Three tests were conducted for each imager. Technical difficulties with the blackbody used as the hot surface prevented the collection of data for all imagers. Mode shifts can be observed as jumps in the data near the same hot surface temperature for all three tests. Imagers 3 and 4 produced higher *P* values after the mode shifts. Imager 5 did not display a pronounced mode shift.

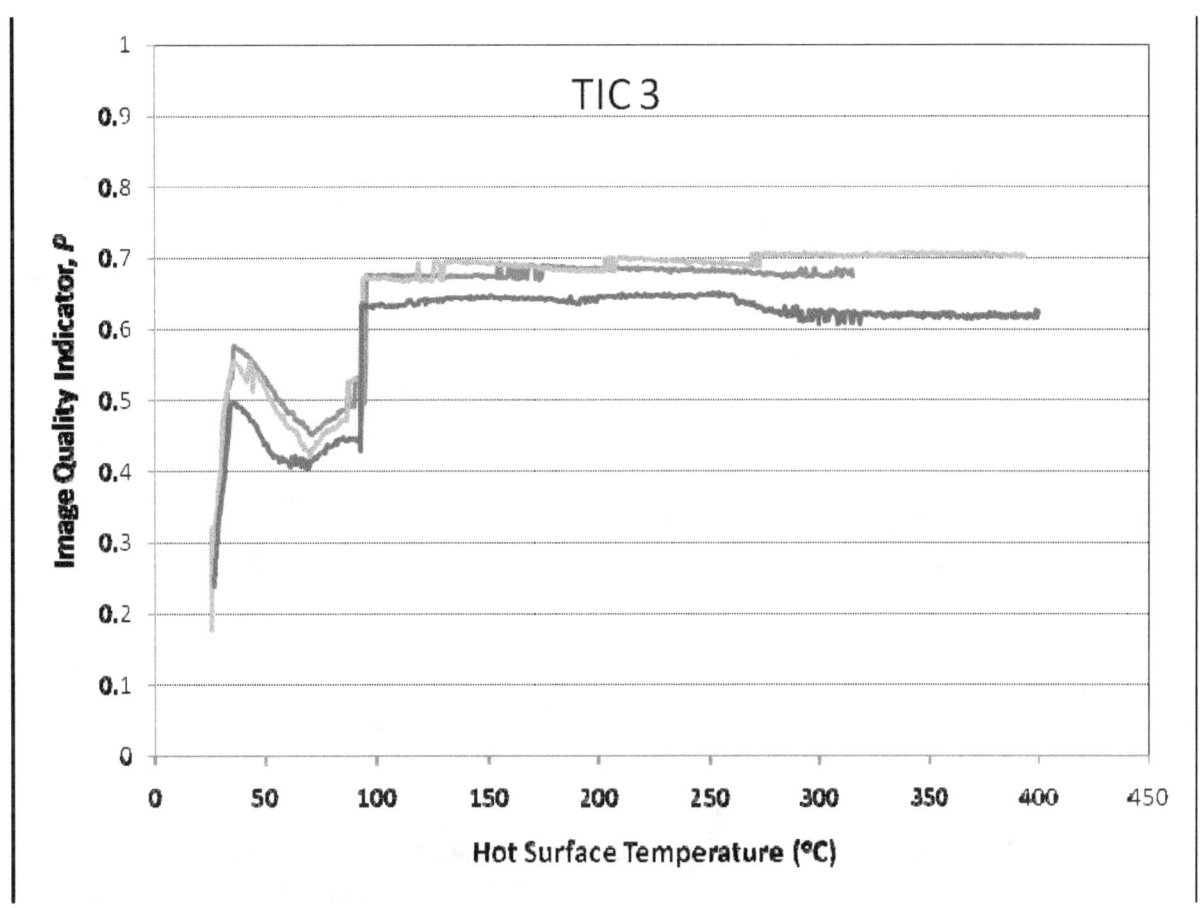

Figure 11. Effective Temperature Range (*ETR*) for Imager 3 plotted in terms of the image quality indicator, *P*. Each colored line represents the results of one test.

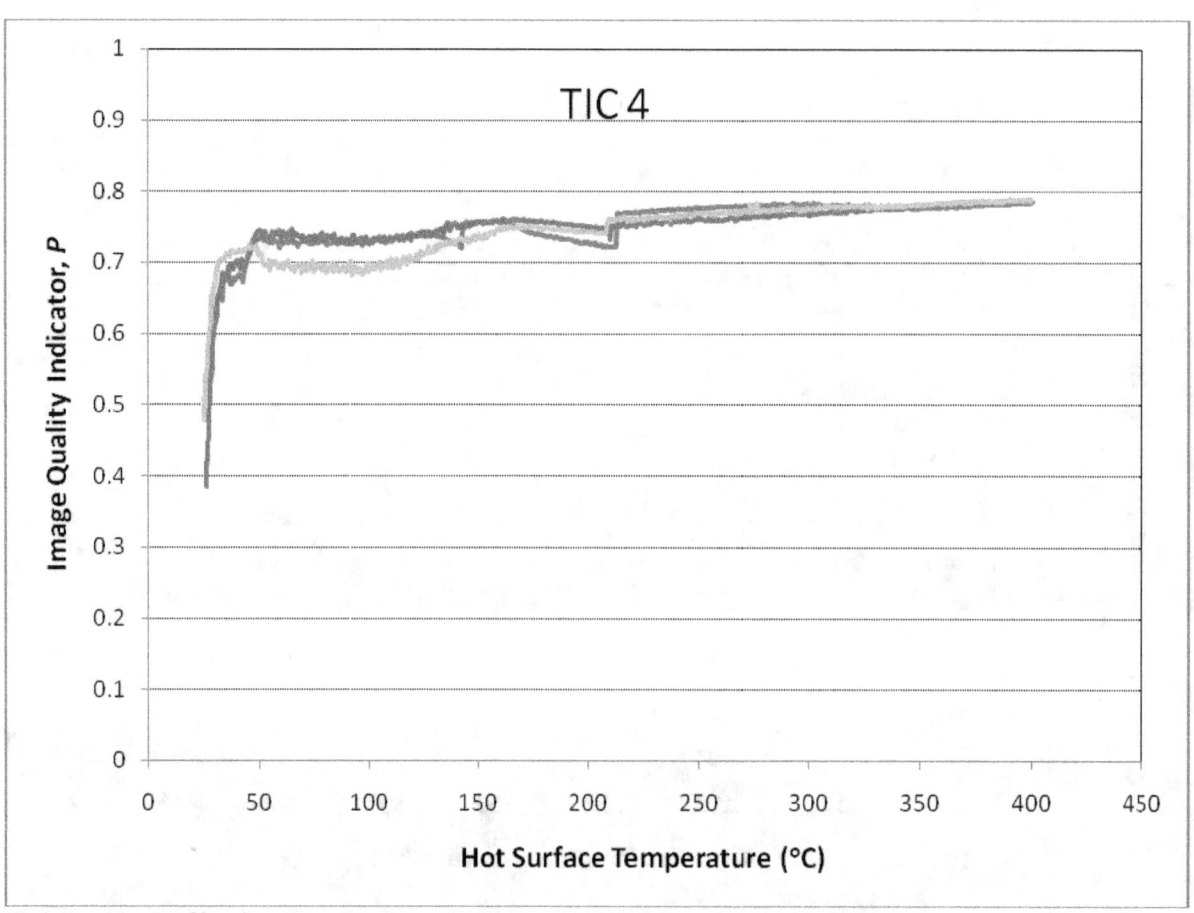

Figure 12. Effective Temperature Range (*ETR*) for Imager 4 plotted in terms of the image quality indicator, *P*. Each colored line represents the results of one test.

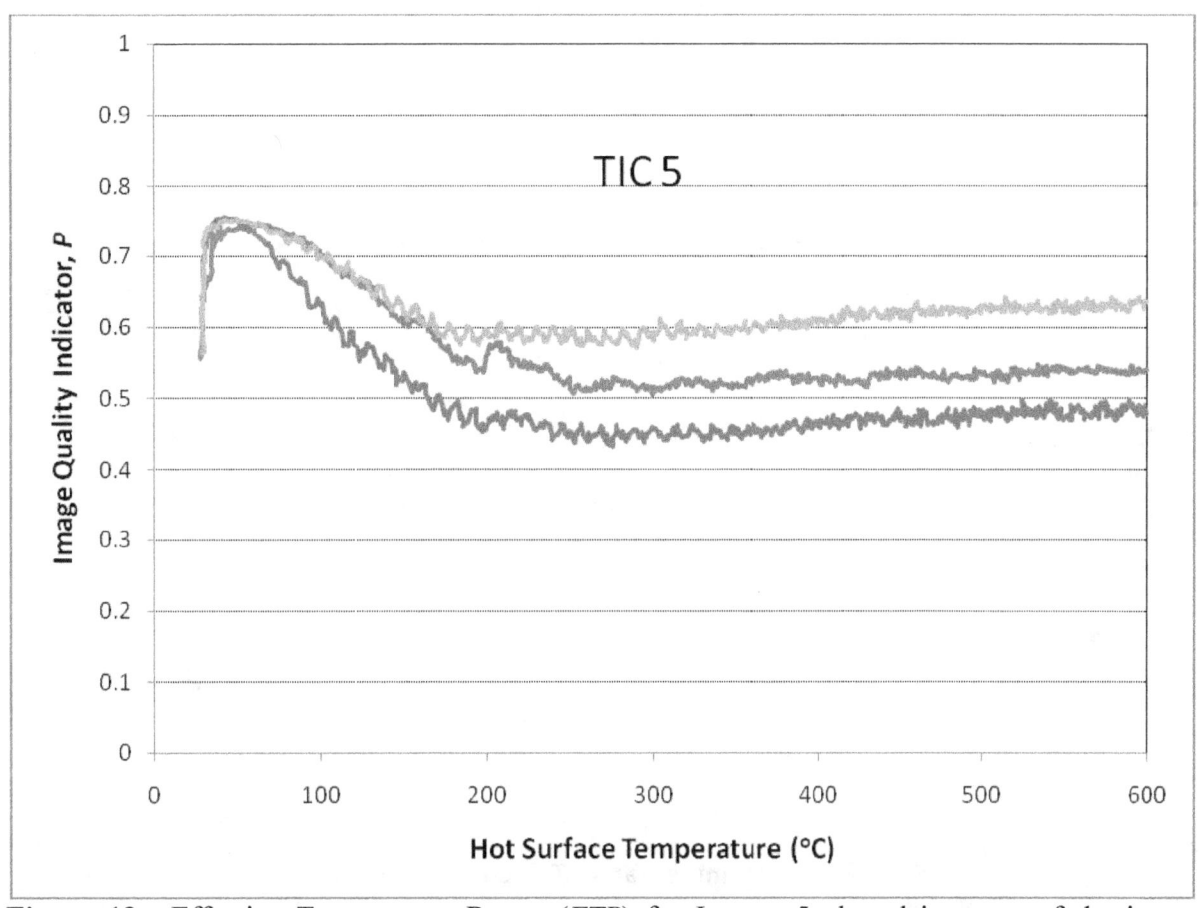

Figure 13. Effective Temperature Range (*ETR*) for Imager 5 plotted in terms of the image quality indicator, *P*. Each colored line represents the results of one test.

3.4.4 Thermal Sensitivity

The thermal sensitivity (*TS*) results, expressed in terms of the image quality indicator, *P*, are shown below in Figure 14. All the imagers appeared to perform best at the cool nominal temperature of 1 °C. There is no significant trend in the test results at nominal target temperatures above 1 °C. These chaotic results indicate the need to look more closely at the temperature difference (T_n) used in the test. Theoretically, if the imager's detector is the limiting system component, the imager should show a decrease in *P* when operating in the low sensitivity mode (shorter integration time), which is not apparent in Figure 14.

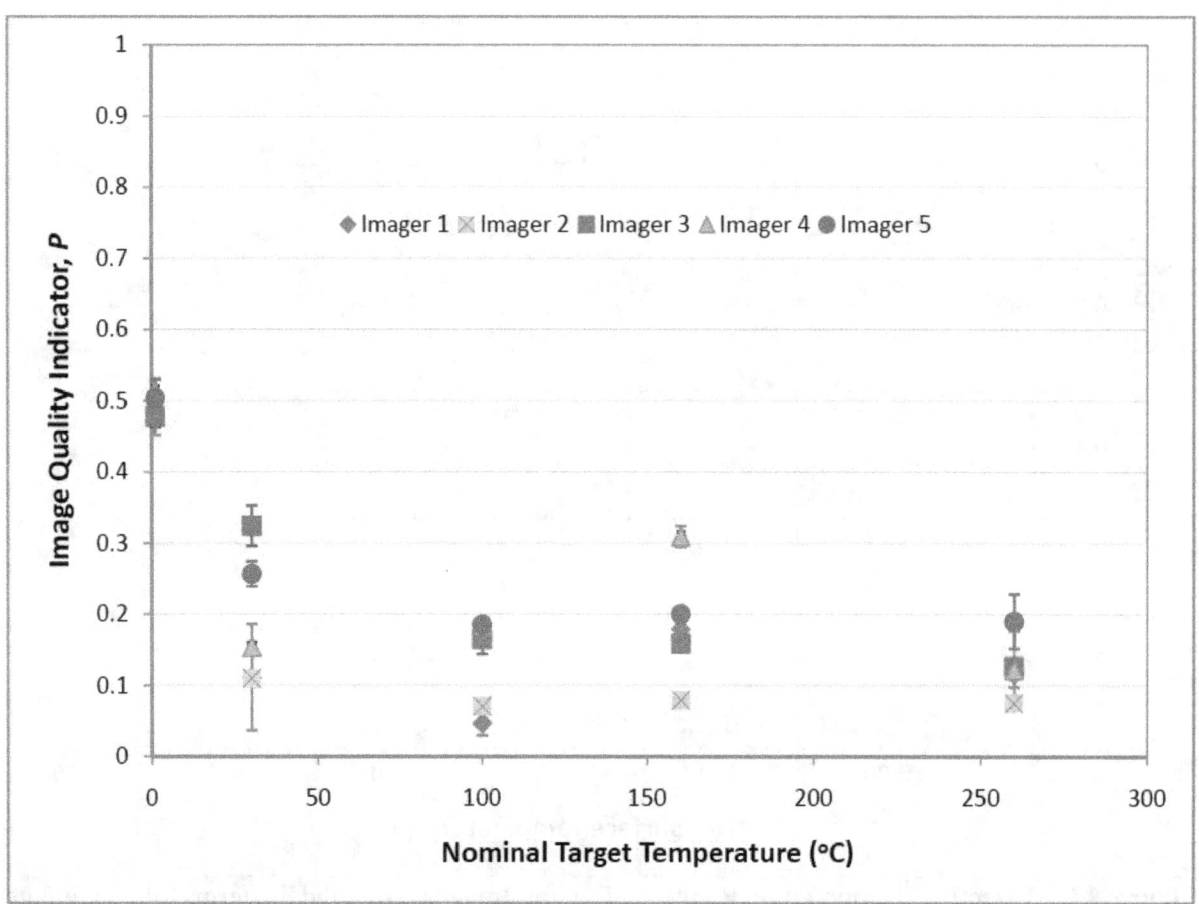

Figure 14. Thermal Sensitivity (*TS*) plotted in terms of the image quality indicator, *P*. The uncertainty expressed in these test measurements is a combination of Type A (statistical) and Type B (other), with a coverage factor of 2 resulting in a 95 % confidence interval. See Section 5 for detailed uncertainty analysis.

There are several factors to consider when attempting to measure very fine differences in temperature. First, the target temperature must be well characterized and measured with sufficient accuracy. Second, the thermal sensitivity of the temperature sensing device under test must be capable of resolving the temperature difference. Third, the displayed image must be resolved with enough shades of gray to provide a contrast that can be perceived by the user.

The blackbody used to produce T_n in this test (CI Systems SR-800) has a temperature accuracy of 8 mK for temperatures below 50 °C and 15 mK for temperatures above 50 °C, however, its temperature range is restricted to 0 °C to 175 °C. For the 260 °C test point, the CI Systems SR-80-7HT has a temperature accuracy of 0.5 °C, although there are other commercially available blackbodies that are more accurate at this temperature. If this test method was modified to require thermocouple temperature measurements, the accuracy of these measurements falls to 1.1 °C or 0.4 %, whichever is greater. A differential blackbody could be utilized, however, these instruments are restricted to temperature differences near ambient temperature.

The thermal sensitivity of the imager's detector comes into play when the dominant temperature difference in the imager's field of view is relatively small. For the nominal target temperature of

30 °C, the limiting component would be the detector if the user could discern individual levels of grayscale.

The displays used for fire service thermal imagers have 8-bit grayscale resolution; therefore, scenes in which temperature differences exist on the order of 250 °C could not be expected to be resolved to less than in 1 °C increments. Following this logic, the smallest value of T_n that can be resolved on the imager's display (one increment of grayscale) for each of the nominal target temperatures, assuming an ambient temperature of 25 °C and a 50 mK detector thermal sensitivity, is listed in Table 8.

Table 8. Minimum resolvable T_n due to imager component limitations.

Nominal Target Temperature (°C)	Minimum T_n	Limitation
1	94 mK	display
30	50 mK	detector
100	293 mK	display
160	527 mK	display
260	918 mK	display

Given the minimum T_n values needed to produce one increment of grayscale on an 8-bit display at the nominal target temperatures required in this test sequence, it is useful to examine the response of the imager to increasing values of T_n. This is done for one imager in Figure 15. The data presented in Figure 15 shows that, when plotted in terms of P, holding all the other image quality factors constant and increasing the T_n from 50 mK to 3000 mK, P remains quite low while the uncertainty in the measurement increases substantially.

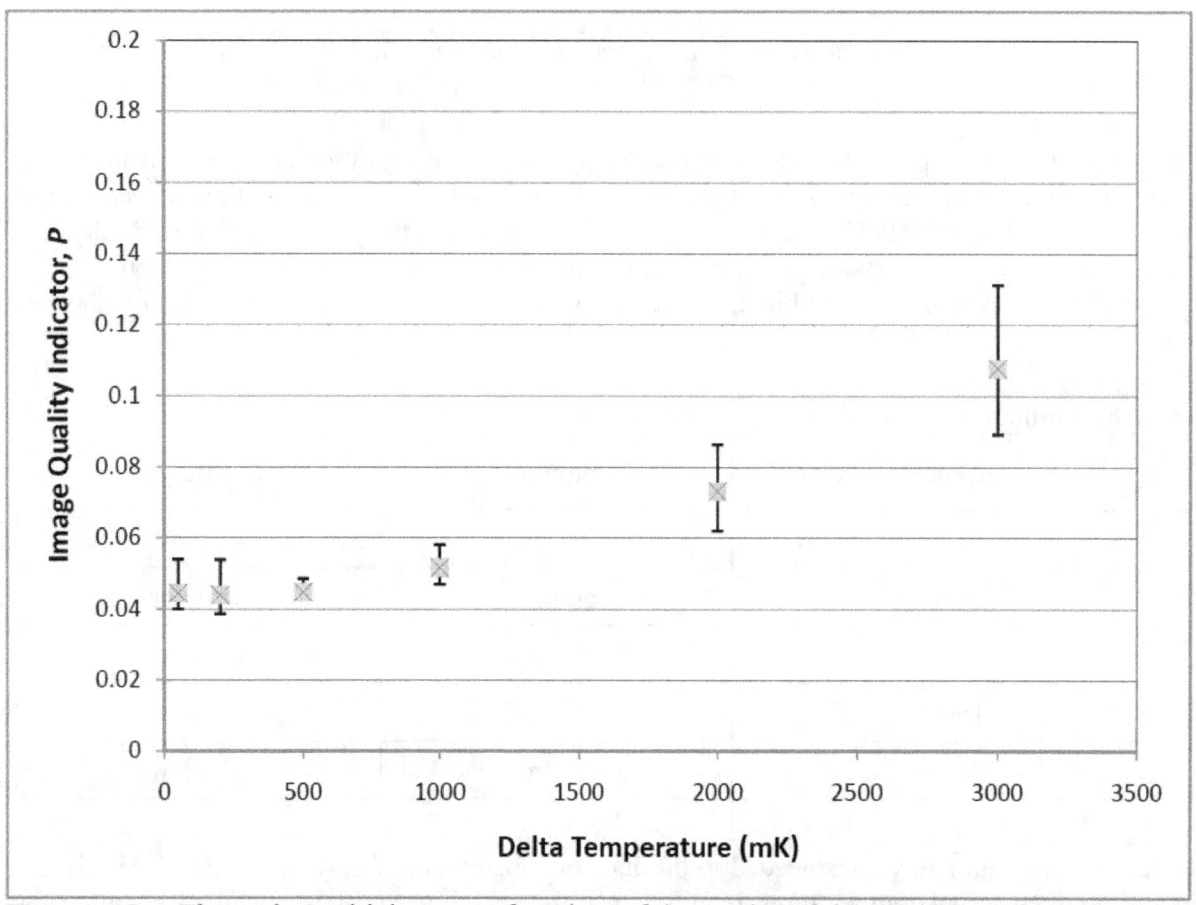

Figure 15. Thermal sensitivity as a function of increasing delta T (T_n). The uncertainty expressed in these test measurements is a combination of Type A (statistical) and Type B (other), with a coverage factor of 2 resulting in a 95 % confidence interval. See Section 5 for detailed uncertainty analysis.

In order to improve the application of this test to fire service thermal imagers, several options may be considered. Possible changes to the test method are listed below, note that this is not an exhaustive list.

1. Leave the test method as is.

2. Make this measurement only at one (ambient) nominal target temperature, using a differential blackbody to measure the temperature difference.

3. Use the imager's detector thermal sensitivity specification, which is measured as the NETD, as the imager's thermal sensitivity.

4. Adjust the T_n values as necessary to account for resolution limitations imposed by system components.

5. Lower the acceptable image quality indicator value, P, to a value that does not eliminate all the thermal imagers in the commercial fire service market.

6. As a last resort, remove this test method from the suite of image quality tests until a more suitable test method is developed.

4. UNCERTAINTY ANALYSIS

There are different components of uncertainty in the measurements made with the equipment used in the tests discussed in this report. Type A uncertainties are those which are evaluated using statistical methods and Type B uncertainties are those which are estimated using other means, such as experience with a particular type of equipment. Type B uncertainties are evaluated by estimating the upper and lower limits for the quantity in question such that the probability that the quantity would fall within the upper and lower limits is essentially 100 %.

After estimating the uncertainties by Type A and Type B analysis, the results are combined in quadrature to yield the combined standard uncertainty. When the combined standard uncertainty is multiplied by a coverage factor of two, the result is the expanded uncertainty which correspond to a 95 % confidence interval (2σ).

The components of uncertainty are listed in Table 9. Some of these components, such as the blackbody temperature measurements, are derived from instrument specifications. Other components, such as the angular *FOV* measurements, are estimated based on estimates of uncertainty in reading the angular markings on the protractor or rotating platform.

Table 9. Components used in uncertainty analysis.

Component	Type A Uncertainty	Type B Uncertainty	Combined Uncertainty	Total Expanded Uncertainty
FOV angle	σ	0.25 degrees	$\sigma + 0.25$ degrees	0.6 degrees
SR-800 Blackbody	σ, pixel intensities in ROI	$T < 50\,^{\circ}C: \pm 8$ mK $T > 50\,^{\circ}C: \pm 15$ mK	$\sigma + 8$ mK $\sigma + 15$ mK	$2(\sigma + 8$ mK$)$ $2(\sigma + 15$ mK$)$
SR-80-7HT Blackbody	σ, pixel intensities in ROI	$\pm 0.5\,^{\circ}C$	$\sigma + 0.5\,^{\circ}C$	$2(\sigma + 0.5\,^{\circ}C)$
IRCon Blackbody	σ, pixel intensities in ROI	$T < 316\,^{\circ}C: 2\,^{\circ}C$ $T > 316\,^{\circ}C: 0.5$ %	$\sigma + 2\,^{\circ}C$ $\sigma + 0.5$ %	$2(\sigma + 2\,^{\circ}C)$ $2(\sigma + 0.5$ %$)$
Type J thermocouples	σ	Greater of 1.1 $^{\circ}C$ or 0.4 %	$\sigma + 1.1\,^{\circ}C$ or $\sigma + 0.4$ %	$2(\sigma + 1.1\,^{\circ}C$ or $\sigma + 0.4$ %$)$
Type K Thermocouples	σ	Greater of 1.1 $^{\circ}C$ or 0.4 %	$\sigma + 1.1\,^{\circ}C$ or $\sigma + 0.4$ %	$2(\sigma + 1.1\,^{\circ}C$ or $\sigma + 0.4$ %$)$

The error bars and measurement accuracies shown in the figures and tables, respectively, in this report reflect values calculated using the uncertainties listed in Table 8 above.

5. CONCLUSIONS

Five thermal imagers were provided by fire service thermal imager manufacturers for evaluation using test methods designed specifically to assess their image quality performance for fire service applications. These tests were Field of View (*FOV*), Nonuniformity (*NU*), Spatial Resolution (*SR*), Effective Temperature Range (*ETR*), and Thermal Sensitivity (*TS*).

The *FOV* test results show that the design of thermal imager optical systems varies significantly, even in cases where the same sensing material and detector array size are used. The combined horizontal and vertical *FOV*s differed by approximately 75 % from the narrowest to the widest *FOV* values.

The *NU* test results, when considered in the context of human perception using the image quality indicator (*P*), were generally reproducible and showed that all the imagers performed in the range of about 0.6 to 0.95 across the set of five target temperatures. The uncertainty in these measurements was larger for the 100 °C and 160 °C target temperatures, which is consistent with observations that some of the imagers shifted sensitivity mode for one or two of the three tests performed at these intermediate target temperatures.

The *ETR* tests show that, after an initial increase from ambient temperature, the three imagers tested gave three distinctly different responses. Imager 3 has a higher value after the mode shift, imager 4's performance didn't change noticeably due to mode shifts and was consistently higher than the other two imagers, and the performance of imager 5 slowly decreased and then reached a plateau with no indication of a mode shift.

The results of the *TS* tests are problematic. As the nominal target temperature increases above 1 °C, the performance of the imagers drops drastically and then becomes erratic at target temperature above 100 °C. Part of the erratic nature of the data collected at higher temperatures is due to the mode shift, for the same reason noted in the *NU* discussion above. This test does not appear to fully capture the actual thermal sensitivity of the imager. A closer examination of the response of two imagers to increasing the temperature difference (T_n) used in this test showed that the imagers don't discriminate significantly between a T_n of 50 mK and a T_n of 2000 mk.

References

[1]   Amon, F., et al., *Performance Metrics for Fire Fighting Thermal Imaging Cameras – Small- and Full-Scale Experiments*. 2008

[2]   Holst, G.C., *Testing and Evaluation of Infrared Imaging Systems*. 1998, Winter Park, FL: JCD Publishing.

[3]   Lock, A. and F. Amon. *Measurement of the Nonuniformity of First Responder Thermal Imaging Cameras*. in *SPIE Defense + Security*. 2008. Orlando, FL: SPIE.

[4]   Donnelly, M.K., et al., *Thermal Environment for Electronic Equipment used by First Responders*. 2006(NIST Technical Note 1474).

[5]   IEC, *Multimedia systems and equipment - Colour measurement and management* International Electrotechnical Commission, 1999. IEC 61966-2-1.

[6]   Lock, A. and F. Amon. *Application of Spatial Frequency Response as a Criterion for Evaluating Thermal Imaging Camera Performance*. in *SPIE Defense and Security Conference*. 2008. Orlando, Florida: SPIE.

[7]   ISO, *Photography - Electronic still-picture cameras - Resolution measurements*. International Standards Organization 2000. ISO 12233.

6. APPENDIX - EQUIPMENT LIST

Certain commercial entities, equipment, or materials may be identified in this document in order to describe an experimental procedure or concept adequately. Such identification is not intended to imply recommendation or endorsement by the National Institute of Standards and Technology, nor is it intended to imply that the entities, materials, or equipment are necessarily the best available for the purpose. When possible, readily available equipment that had already been purchased for different tasks was utilized rather than purchasing new equipment that might be better suited for the specific tasks in this project.

1. Nikon D3 digital SLR camera.
   http://www.nikonusa.com/Find-Your-Nikon/ProductDetail.page?pid=25434

   Remote control cord.
   http://www.nikonusa.com/Find-Your-Nikon/Product/Remote-Cords/4652/MC-22-Remote-Cord-with-Banana-Plugs.html

   Nikkor 28 mm lens.
   http://www.nikonusa.com/Find-Your-Nikon/Product/Camera-Lenses/1922/AF-NIKKOR-28mm-f%252F2.8D.html

   +10 close-up filter.
   http://www.bhphotovideo.com/c/product/94232-REG/Hoya_S52MCU_52mm_Macro_Close_up_10.html

   Image conversion software (apparently the newest Photoshop CS4 includes the plug-in necessary for use with Nikon D3 files).
   http://bibblelabs.com/ or http://www.adobe.com/products/photoshop/cameraraw.html

   Memory.
   http://www.newegg.com/Product/Product.aspx?Item=N82E16820211244

   Camera stand.
   http://www.bhphotovideo.com/bnh/controller/home?ci=0&shs=Manfrotto+by+Bogen+Imaging+SALON+230+CAMERA+STAND+(7%27)+-+BO809&sb=ps&pn=1&sq=desc&InitialSearch=yes&O=jsp%2Fcatalog.jsp&A=search&Q=*&bhs=t&Go.x=0&Go.y=0&Go=submit

   Articulated camera positioning arm.
   http://www.bhphotovideo.com/bnh/controller/home?O=search&A=search&Q=&shs=manfrotto+by+bogen+imaging+articulated+arm+bracket+bo396b3&ci=0

   Grip kit for the articulating arm.
   http://www.bhphotovideo.com/bnh/controller/home?ci=0&shs=Avenger+GRIP+KIT%3A+C1575B%2FD200B%2FE600%2FD520B%2FBAG+-+AVD800KIT&sb=ps&pn=1&sq=desc&InitialSearch=yes&O=jsp%2Fproductlist.jsp&A=search&Q=*&bhs=t&Go.x=24&Go.y=11&Go=submit

2. Data Acquisition System.

   Controller card.
   http://sine.ni.com/nips/cds/view/p/lang/en/nid/14235

Chassis.
http://sine.ni.com/nips/cds/view/p/lang/en/nid/10676

Analog In module, terminal block, and cable (use TC-2095 terminal).
http://sine.ni.com/nips/cds/view/p/lang/en/nid/1654
http://sine.ni.com/nips/cds/view/p/lang/en/nid/1676
http://sine.ni.com/nips/cds/view/p/lang/en/nid/1834

Analog Out module and terminal block. This is to control relays that control the Nikon D3 camera.
http://sine.ni.com/nips/cds/view/p/lang/en/nid/1870
http://sine.ni.com/nips/cds/view/p/lang/en/nid/1682

3. Emissivity paint.
Optical black coating, P/N 20825, Medtherm Corp., P. O. Box 412, Huntsville, AL.
Average emissivity = 0.95, ± 0.03.

4. Thermocouples.
Use Type-J thermocouples.
http://www.omega.com/pptst/SA1.html

5. Spatial resolution target.
See files: ImRecTarget.svg, ImRecTarget.eps, ImRecTarget.ai. Note- these are all the same image but with different file formats.

The 24" x 36" silicone rubber heating blanket for the background comes from:
http://www.briskheat.com/p-347-srl-srp-silicone-rubber-heating-blankets.aspx (use a variac to control the temperature)

Metal target stencil.
Frederick Sign and Banner, 18 East 6th St., Frederick, MD 21701, 301-663-9122.

6. Blackbodies.
One extended area blackbody that can go as low as 0 $^{\circ}$C, such as one of these:
http://www.ci-systems.com/htmls/article.aspx?C2004=12724&BSP=12547 (4", extended temperature range).

http://www.sbir.com/blackbody_13132.htm (4", -30 to 100 $^{\circ}$C temperature range)

http://www.mikroninfrared.com/Catalog.aspx?id=1164&ekmensel=c580fa7b_8_28_1150_2
(4", -5 to 170 $^{\circ}$C temperature range)

http://www.electro-optical.com/html/datashts/extended/pdf/eo710ces200.pdf
(4", note: temperature accuracy might be out of specification for NFPA 1801)

In all the above cases, the nonuniformity of the blackbody as measured in paragraph 8.1.5.13 needs to be verified.

A high temperature extended area blackbody.
http://www.ci-systems.com/Htmls/article.aspx?C2004=12726&BSP=12547
(Temperature range 30 $^{\circ}$C to 550 $^{\circ}$C)

http://www.sbir.com/blackbody_4000.htm

(Temperature range 30 $^{\circ}$C to 550 $^{\circ}$C)

http://www.mikroninfrared.com/Catalog.aspx?id=2392&ekmensel=c580fa7b_8_28_1150_2

http://www.electro-optical.com/html/datashts/extended/pdf/CES600800RevisedC.pdf
(Temperature range 30 $^{\circ}$C to 550 $^{\circ}$C)

7. Effective Temperature Range Bars.

    4 bars: 1.3 cm (½ in) thick and set 1.3 cm (½ in) apart; they are 15.2 cm (6 in) long. They must maintain a constant temperature of 30 $^{\circ}$C ± 0.5 $^{\circ}$C. Suggestion: use thermoelectric (peltier) heaters with a variac to control them.